Ana Beatriz Ramos M. Abrahão
Jonas Frank Reis
Fábio H. C. Bernardes

Development of oxyacetylene welding for advanced materials

Ana Beatriz Ramos M. Abrahão
Jonas Frank Reis
Fábio H. C. Bernardes

Development of oxyacetylene welding for advanced materials

Aluminium alloy 2024 and PEI/Fibreglass composite

ScienciaScripts

Imprint

Cover image: www.ingimage.com

This book is a translation from the original published under ISBN 978-613-9-69216-3.

Publisher:
Sciencia Scripts
is a trademark of
Dodo Books Indian Ocean Ltd. and OmniScriptum S.R.L publishing group

120 High Road, East Finchley, London, N2 9ED, United Kingdom
Str. Armeneasca 28/1, office 1, Chisinau MD-2012, Republic of Moldova, Europe
Printed at: see last page
ISBN: 978-620-8-12510-3

ACKNOWLEDGEMENT

I would firstly like to thank God, my parents, grandparents and family for their support. Especially to my grandfather Vantuir Bernardes (In Memory).

My special thanks go to Prof.ª Dr.ª Ana Beatriz R. M. Abrahão, who trusted and believed in my potential to develop this degree work with quality and efficiency.

And to all the teachers and classmates who directly or indirectly played a part in my education.

"If I could see further, it was by standing on the shoulders of giants" (Isaac Newton).

Dreams determine what you want. Actions determine what you achieve.

(Aldo Novak)

SUMMARY

Aluminium is the second most widely used metal in the world, second only to steel. This is due to the abundance of its raw materials on earth and the possibility of forming important alloys, such as the 2000 series alloy, AA 2024, which has been studied extensively due to its applicability in the aerospace industry where materials are needed that combine high mechanical strength with low density. In order for this alloy to be used, efficient adhesion with dissimilar materials must occur, making it possible to combine attributes and obtain greater efficiency in the application.

The joining of these materials is still significantly complex, but it gives them greater flexibility and specific localised properties, allowing excellent characteristics for aircraft components, reducing weight and increasing performance.

Among these dissimilar materials, composites are most commonly used in the aeronautics industry, as they have characteristics that, in addition to gaining properties, reduce weight by around 25% and allow for a gain in performance compared to metallic structures already in use. Amongst these, polymer composites (polyimides) are standing out, in which PEI (polyetherimide) has received the most research attention due to its low cost, easy handling and the fact that it shows superior results compared to other polymers in numerous tests such as fatigue, traction and impact.

The aim of this undergraduate project is to join the AA2024 aluminium alloy with PEI in a pioneering way using the oxyacetylene welding process with an innovative methodology that enables experimentation.

Keywords: Oxyacetylene welding. AA2024 aluminium. PEI-Fibreglass. Dissimilar materials.

SUMMARY

CHAPTER 1

INTRODUCTION

Aluminium is the second most widely used metal in the world, second only to steel. Aluminium can form important alloys that have properties such as ductility, good thermal and electrical conductivity, resistance to corrosion in a less aggressive environment, all combined with the lightness of the material, as well as being suitable for numerous subsequent processes, such as rolling, forging, extrusion, drawing, machining, welding, among others (ABAL, 2007; KHAN, 2011).

One of the alloys that has been studied the most recently is the 2XXX AA 2024 series alloy. This is due to its applicability in the aerospace industry where materials are needed that combine high mechanical strength with low density. Containing copper, magnesium, manganese and other elements in smaller quantities, this alloy is generally worked by extrusion and hot rolling. As well as having good mechanical strength and low density, other factors that make this alloy very common are that it can be aged naturally or artificially and can be hardened by second phase precipitation, which can surpass even medium carbon structural steel (COUTO et al., 2012).

In order to utilise the AA2024 alloy in the aeronautical field, it is necessary for this alloy to adhere efficiently to dissimilar materials, making it possible to reconcile attributes and obtain greater efficiency in the application. The joining of dissimilar materials is usually done using a metallic material with a composite, each with different possible properties and usually made up of long fibres that allow these properties to be distributed more evenly throughout the material (ZANATTA, 2012).

As a result of these benefits, thermoplastics are being more widely used, studied and researched. Amongst these, polymer composites (polyimides) are standing out, in which PEI poly(ether-imide) has received the most research attention due to its low cost, easy handling and because it shows superior results compared to other polymers in numerous tests such as fatigue, traction and impact (BOTELHO, 2007).

Among the aluminium alloys, the 2XXX series, particularly the 2024 T3, is used extensively in the aeronautical industry due to its high mechanical strength, greater hardness and tensile strength than aluminium, and also its high specific strength (strength to weight ratio) compared to other light alloys (ABAL, 2007; KHAN, 2011).

The development of hybrid composite technology, combining polymer laminates reinforced with long fibres and metal plates, is aimed at forming a set of materials that combine high mechanical strength and stiffness values with low specific mass. Currently, mechanical fixation and adhesive bonding are the most widely applied joining techniques for metal/composite hybrid structures. However, due to the limitations of these techniques, such as the curing time of thermo-rigid adhesives and the formation of stress concentration by mechanical fastening, new welding-based technologies have been developed Oxy-gas welding is a fusion welding process in which the union between metals is achieved through the application of heat generated by one or more flames resulting from the combustion of a gas, with or without the aid of pressure, and there may or may not be filler metal. The system is simple, consisting of compressed gas cylinders, pressure regulators, pressure gauges, hoses, check valves and a welding torch with a suitable nozzle; different atmospheres can be achieved by varying the relative amount of oxidiser and fuel. The ability of metals to bond with polymers or composites is indeed a major challenge (ABRAHÃO, 2018; ANDRÉ, 2016).

The aim of this undergraduate project is to join the AA2024 aluminium alloy with PEI in a pioneering way using the oxyacetylene welding process with an innovative methodology that makes it possible to join dissimilar materials even though it is still significantly complex, but makes it possible to give the project greater flexibility and specific localised properties, allowing excellent specific characteristics for aircraft components, reducing weight and increasing their performance.

CHAPTER 2

LITERATURE REVIEW

2.1. OXY-GAS WELDING (OFW)

The oxy-gas welding process uses the heat generated by burning a combustible gas with an oxidising gas (oxygen), generating coalescence between the materials to be welded. The procedure can be autogenous or with filler metal, which when used melts and solidifies together with the fusion pool. Other possibilities for this process, depending on the flame setting, are to be used for pre- and post-heating, oxy-cutting, brazing and even to heat-treat small parts (MODENESI, et.al 2011).

As well as being a versatile process and having portable equipment that is easy to transport, other advantages that this process offers the welder are: control of the heat and temperature of the parts, bead size, shape and viscosity of the molten pool, which is possible since the feed of the filler metal and control of the heat source are achieved independently. A disadvantage of the process is that it is not economically viable to weld very thick parts, except in repair situations (BRACARENSE, 2000).

To carry out oxy-gas welding, you basically need a torch, which mixes and controls the flow of gases at the nozzle outlet, which are usually made of copper or some other material that has good thermal conductivity characteristics to prevent overheating and allow easy heat transfer from the flame to the molten pool. The oxygen and fuel gas cylinders have pressure regulators to control the cylinder's internal pressure and keep the flow rate stable, allowing for better welding conditions. The welding equipment also has safety valves, which are indispensable because they reduce risks and even prevent serious accidents. The valves are flame backflow and counterflow valves, whose function is to block the counterflow of gases and eliminate flame backflow (BRANDI et.al 1992).

During the soldering operation, some materials are consumable, such as the gases, the filler metal and the flux (when used). The gases used are oxygen and a combustible gas, usually acetylene. The cylinders can be connected to an industrial plant when used on a large scale, or they can be in small cylinders on a portable trolley (MODENESI, et.al 2011).

Oxygen is a combustible, colourless, insipid gas that accelerates the reactions that take place in the air and can react violently with oil or grease. For this reason, cylinders (including those of other gases) should be stored without lubrication, in a clean, ventilated environment and out of contact with electrical cables and conductors. Acetylene is a combustible gas, colourless in its gaseous state and odourless when pure. Under normal conditions it is lighter than air and combusts very easily. Fluxes are generally only used for soldering non-ferrous metals such as aluminium, copper and its alloys and cast iron. Steel in general does not need flux. These improve the fluidity of the molten pool and wettability, as well as reacting with metal oxides present on the surface and forming a layer of protective slag (MODENESI, et.al 2011).

Oxy-acetylene welding can be carried out in two ways: from left to right (backhand), where the flame is directed towards the weld bead; and from right to left (forehand), where the flame is pointed in front of the weld bead, as if it were pushing the weld pool. The forehand technique is suitable for welding sheets up to 3mm thick, while the backhand technique gives better results with thicker sheets. This is because when welding from the right, the flame is further away from the weld pool (BRANDI et.al 1992).

In addition to these two types of execution, there can be a combination with a transverse movement (weaving), which allows greater control of the melt pool and enables the formation of wider beads and greater penetration in bevelled parts (MODENESI, et.al 2011).

To make a quality weld, it is important to regulate the flame properly. When it opens, the flame is acetylenic (very sooty) and gradually the oxygen valve must be opened to regulate the size and type of flame suitable for welding. The flames can be of the reducing, neutral or oxidising type as shown in Figure 1 and are obtained successively in this order as the oxygen flow rate is increased (MODENESI, et.al 2011).

Regulagem da chama	Tipo da chama	Formato da chama	Característica	Aplicação
a < 1,0	Redutora		Penacho esverdeado. Véu branco circundando o dardo. Dardo branco, brilhante e arredondado. Chama menos quente.	Revestimento duro, ferro fundido, alumínio e chumbo
1,0 < a < 1,1	Neutra		Penacho longo. Dardo branco, brilhante e arredondado.	Soldagem de aços (ou regulagem neutra levemente redutora). Cobre e suas ligas (exceto latão). Níquel e suas ligas.
a > 1,1	Oxidante		Penacho azulado ou avermelhado, mais curto e turbulento. Dardo branco, brilhante, pequeno e pontiagudo. Chama mais quente. Ruído característico.	Aços galvanizados (regulagem neutra levemente oxidante) Latão Bronze

Figure 1- Types and Characteristics of Flames.

Source: (Adapted from BRANDI et.al 1992).

Due to the ease and versatility of this process, it can be easily applied to ferrous and non-ferrous metals and alloys, such as low, medium and high carbon steels, aluminium alloys, copper, among other alloys. When welding non-ferrous alloys, it is important to adapt the process according to the characteristics of the material, depending mainly on the melting point and thermal conductivity, and flux and supports can be used to hold the part and maintain the welding angle (BRACARENSE, 2000).

Despite being simple and versatile, this welding process is not very common in industry due to its low productivity and is generally used to weld non-ferrous materials or other materials in repair situations (MODENESI, et.al 2011).

2.2. AA2024 ALUMINIUM ALLOY

Aluminium is the second most widely used metal in the world, second only to steel. This is due to the abundance of its raw material in the earth (bauxite) and its properties such as ductility, good thermal and electrical conductivity, resistance to corrosion in a less aggressive environment, all combined with the lightness of the material, which has just over a third of the density of steel. Furthermore, although obtaining pure aluminium is expensive due to the electrolytic reduction process, the end product is suitable for numerous subsequent processes, such as rolling, forging, extrusion, drawing, machining, welding, among others (ABAL, 2007; KHAN, 2011).

Although pure aluminium has many qualities, it is limited when it comes to mechanical strength. In an attempt to solve this shortcoming, in 1903 the German Alfred Wilm discovered that it was possible to harden aluminium by forming an alloy with around 4% copper, causing precipitation hardening. Small amounts of magnesium and manganese were later added to this alloy, which gave rise to the 2XXX series alloys (ABREU, 2016).

Of the 2XXX series, the AA 2024 alloy is the one that has been most studied recently due to its applicability in the aerospace industry where materials are needed that combine high mechanical strength with low density. Containing copper, magnesium, manganese and other elements in smaller quantities, this alloy is generally worked by extrusion and hot rolling. The addition of copper gradually increases the mechanical strength due to its precipitation, as well as having a low density. Other factors that make this alloy very common are that it can be aged naturally or artificially and can be hardened by second phase precipitation, which can surpass even a medium carbon structural steel (COUTO et al., 2012).

2.3. POLYMER COMPOSITES

Polymer composites made a strong appearance on the market around 1940 during the Second World War, where glass fibres with polyester resin were used in the manufacture of secondary structures for American aircraft due to the reduction in cost, weight, resistance to the corrosive environment (increasing durability) and ease of use.

(MARTINS, 2007). In other branches of industry, composite materials came about to remedy the lack of materials that met the needs of several different properties that were not possible in a single type of material (GOMES, 2015).

In general, composite materials can be defined as a multiphase material, i.e. combining two or more materials with microscopically different shapes, compositions and properties, which cannot be soluble with each other to enable the combination of different properties of the materials that make up the phases of the composites, these being the matrix and the dispersed phases (MARTINS, 2007).

Metal, ceramic or polymeric matrices can be used to form a composite material, as well as other natural composites such as wood, bone, etc. The matrix is usually continuous and surrounds the dispersed phase, and there needs to be affinity between the joined materials so that they respond to external and internal stresses efficiently (DA SILVA, 2014).

The dispersed phase can be made up of unlimited possibilities, but the shapes are restricted to those of fibres (natural or artificial), particles, plates, wool, etc. This phase determines the internal structure of the material to be formed. The matrix, on the other hand, serves to unite the constituents and give the microscopic interface its final shape (SILVA, 2010).

For the adhesion of the matrix to the dispersed phase to be established, there must be interaction between the materials through covalent bonds, hydrogen bonds, van der Walls bonds or electrostatic interaction, in addition to the chemical affinity between the two phases. Normally, the dispersed phase is hydrophilic in nature and the polymeric matrix is hydrophobic. Compatibility can be improved by chemically modifying the surface of one of the components (DA SILVA, 2014).

Composite materials have come to meet the need for equipment that requires unusual properties or a combination of properties impossible for a single material and have been applied in the aeronautical industry, automotive, orthopaedic implants, dentistry, Formula 1, marine platforms, mechanical and civil engineering, etc. Although it is a technology that is still developing numerous new possibilities and has a relatively high cost, the advantages already mentioned compensate for the high investments in research and development of new manufacturing methods or even new types of composites that can meet new working environments (GOMES, 2015; DA SILVA, 2014).

2.4. PEI POLY (ETHER-IMIDE)

Among the composite materials most commonly used in the aeronautical industry, advanced thermoplastics have some advantages over conventional thermosets, as they have better impact resistance, can be used at higher temperatures, are easy to repair, easy to transport and low cost, thus characterising an excellent cost-benefit ratio. As well as thermoplastics having superior resilience, toughness and mechanical resistance to the thermosets already being used in the aerospace industry, they also include the fact that they do not stiffen permanently and can be reheated and moulded countless times (BATISTA; BOTELHO, 2009; BOTELHO; REZENDE, 2012).

As a result of these benefits, thermoplastics are being more widely used, studied and researched. Among them, polymer composites (polyimides) are standing out, in which PEI poly(ether-imide) has received greater research attention due to its low cost, easy handling and because it shows superior results compared to other polymers in numerous tests such as fatigue, traction and impact (BOTELHO; REZENDE, 2012).

PEI is an engineering polymer developed in the mid-1970s, although it has been used since 1982. It has an amorphous structure, a high glass transition temperature, high mechanical strength and rigidity both at room temperature and at high temperatures (up to 190° C), inherent flame resistance with low smoke dispersion, as well as being an insulating material with broad chemical resistance (ABRAHÃO, 2015).

The structure of PEI is made up of repeated aromatic imides and ether units. The rigidly grouped imide molecules with aromatic rings linked by one or two unbranched atoms mean that the bonding energy is high, making the PEI structure very flexible even when heat is applied, which makes this structure little affected by thermal cycles, facilitating its remelting and large-scale production, as well as enabling permanent bonding using heat (welding). This all contributes to excellent dimensional/structural stability and high isotropic mechanical properties when compared to other amorphous or semi-crystalline engineering polymers (ABRAHÃO, 2015; BOTELHO; REZENDE, 2012).

With a vast repertoire of excellent properties and characteristics, poly(ether-imide) has a range of possibilities in industry in general, as well as in the aeronautical industry. It has become a good choice as a matrix for composites that will be applied to aircraft, mainly in internal, secondary and tertiary structures (ABRAHÃO, 2015).

PEI's chemical structure consists of repeated aromatic imides and ether units (Figure 3). Unlike other engineering polymers, PEI can be easily melted and produced on a large scale. PEI's amorphous structure is not greatly affected by thermal processing cycles, contributing to excellent dimensional stability and high isotropic mechanical properties compared to other amorphous or semi-crystalline polymers (OLIVEIRA; GUIMARÃES; BOTELHO, 2009).

The presence of aromatic rings in the PEI structure is an important characteristic of these high-performance polymers. These rings are linked by one or two atoms and there are no paraffinyl groups or branches, which means that the binding energy of the structure is high. In addition, the high Tg (around 210°C) allows PEI to be used in applications that require high temperatures (up to 150°C) and the flexibility of its molecule provides good softening during processing, as well as good chemical resistance (CATSMAN, 2005; ADHIKARI; MAJUMDAR, 2004; NETO, 2009).

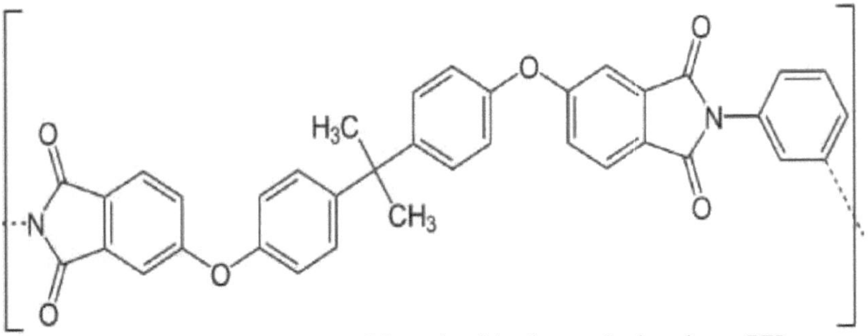

Figure 2- Structural formula of the thermoplastic polymer PEI.

Source: (BARROS, 2007).

PEI is conventionally obtained by the polymerisation process through the condensation of diamines and dianhydrides, and is defined as a high performance amorphous thermoplastic polymer in terms of its properties. These properties include: high chemical resistance to fuels; diluted bases; salt solutions; alcohols; mineral acids and aromatic hydrocarbons. Despite being resistant to these substances, it can be attacked by strong bases and partially halogenated compounds. Other properties include high thermal resistance, excellent modulus of elasticity and tensile strength at high temperatures, excellent dimensional stability, low creep and low coefficient of thermal expansion (SOUZA, 2010).

2.5. Welding thermoplastic materials

The inclusion of joints in structural composites is necessary due to the inherent limitations in the manufacturing process and the importance of these materials in inspection, repair and assembly. Fusion or welding joining methods are promising technologies that have been used in industries dedicated to processing thermoplastic composites, as the efficiency of joining welded parts results in higher quality joints in a shorter time compared to other available techniques (AGEORGES; YE, 2001). All the techniques available for welding thermoplastic composites are adaptable to automation for online inspection. In addition, these techniques provide reproducibility and require minimal surface preparation, thus reducing the cost of the process. Therefore, welding is a very attractive process for joining thermoplastic matrix composites, as it is a fast and simple process in which the parts are joined by fusion and consolidation in the interfacial region of the parts to be welded. Fusion joining techniques can be classified according to the technology used to generate the heat. Figure 6 shows the main fusion joining techniques in which the heating processes are generally classified as thermal welding, friction welding and electromagnetic welding (AGEORGES; YE; HOU, 2000; YOSEFPOUR; HOJJATI; IMMARIGEON, 2004; WISE, 1999).

The joining of certain parts of a component is achieved by bringing the atoms or molecules of the structures to be joined closer together. The principles of bonding thermoplastic components using techniques classified as fusion or welding methods basically consist of preparing the surface to be welded, heating the polymer in the interfacial region to the viscous state that physically causes the interdiffusion of the polymer chains and cooling the polymer, thus obtaining a consolidated bond (MARQUES; MODENESI; BRACARENSE, 2005; DUBÉ et al., 2008). Figure 3 shows the main stages involved in the welding process of a thermoplastic composite.

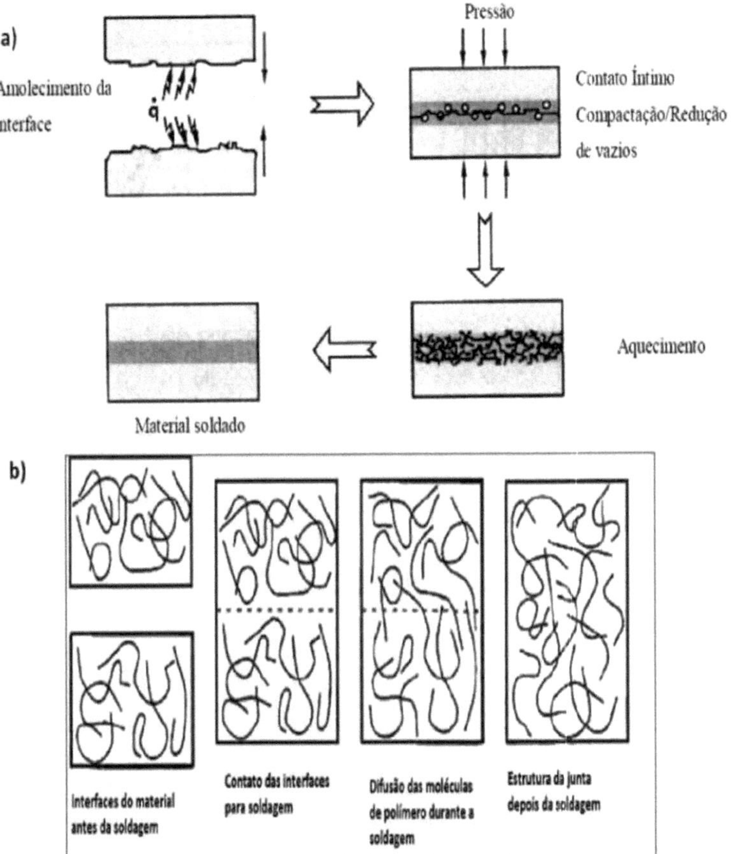

Figure 3 - Joining steps during the welding process of a thermoplastic composite: a) illustration of the samples and b) illustration of matrix behaviour

Source: (DA COSTA, 2011).

Among the various methods available for welding composites, some techniques already stand out, but there is still a lack of more in-depth studies, especially on the feasibility of industrial application. Among the techniques available for integrating composites, fusion joining or welding is the method with the greatest potential for application in assembling, joining and repairing components made of these materials. The main advantages associated with the fusion or welding process compared to the other techniques mentioned above include reduced processing time, reduced sample preparation and low irregularity at the material interface. The main problems related to other composite joining techniques are the need for long processing times and the presence of stress concentration areas in the joined parts (AGEORGES; YE; HOU, 2000; BATES, 2005; DUBÉ et al., 2007; PANNEERSELVAM et al, 2012).

All the techniques available for welding thermoplastic composites are adaptable to automation for online inspection. In addition, these techniques provide reproducibility and require minimal surface preparation, thus reducing the cost of the process. Therefore, welding is a very attractive process for joining thermoplastic matrix composites, as it is a fast and simple process in which the parts are joined by fusion and consolidation in the interfacial region of the parts to be welded. Fusion joining techniques can be categorised according to the technology used to generate the heat. Figure 4 shows the main fusion joining techniques in which the heating processes are generally classified as thermal welding, friction welding and electromagnetic welding (AGEORGES; YE; HOU, 2000; BATES, 2005; WISE, 1999).

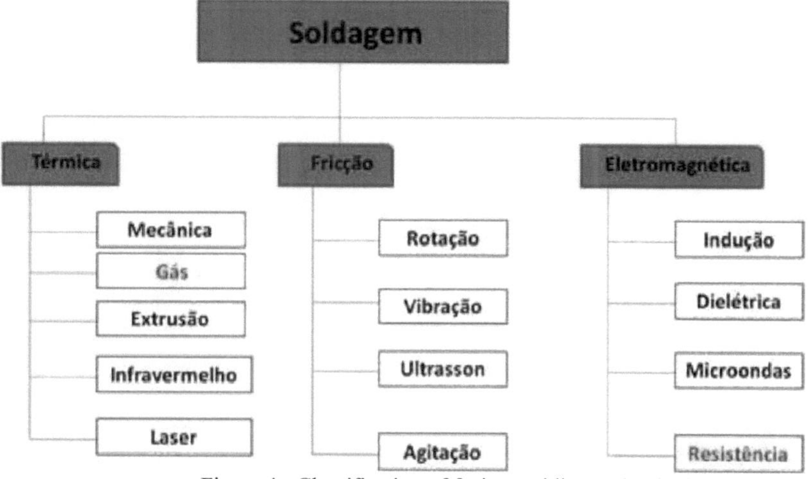

Figure 4 - Classification of fusion welding technologies.
Source: (YOUSEFPOUR; HOJJATI; IMMARIGEON, 2004).

Electrical resistance welding applied to thermoplastic composites has great potential for the integration of high-performance thermoplastic composites, especially when it comes to aerospace applications. However, parameters such as thermal insulation, energy input, welding time, fibre orientation, and the type of resistive element to be used have been studied to improve the quality and performance of welded joints (DA COSTA, 2012;DUBÉ et al, 2007, STAVROV and BERSEE, 2005).

Welding using the oxyacetylene or oxy-gas welding (OFW) process is a process in which the coalescence or joining of materials takes place by means of a flame, which may or may not contain an addition material, with or without pressure. The biggest advantage of applying this technique is that it is a more accessible process, consisting of a low-cost portable system (WAINER; BRANDI; MELLO, 1992; MARQUES; MODENESI; BRACAENSE, 2007).

CHAPTER 3

MATERIALS AND METHODS

3.1. Materials

3.1.1. Laminates

The PEI/glass fibre laminates were made with fabrics in the 8 HS configuration with nominal thicknesses of between 2 and 3.5mm, and in the (0/90) 5s configuration containing approximately 50% by volume of the matrix. These were supplied by TEN-CATE Advanced Composites, as shown in Figure 5.

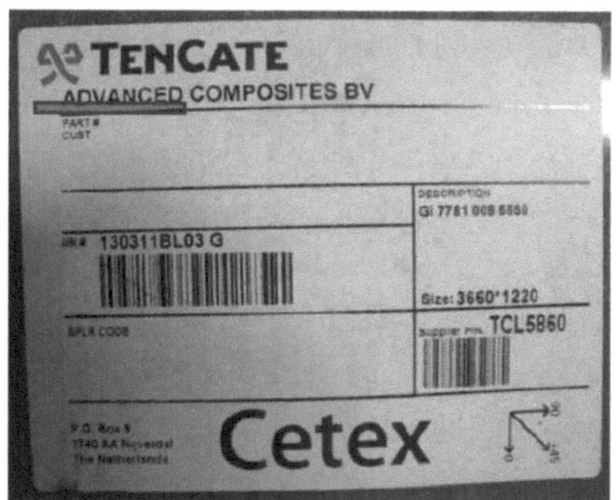

Figure 5- Identification code of the PEI/glass fibre laminate supplier

Source: the Author

3.1.1 Aluminium alloy AA 2024 T3

The 2024 T3 aluminium alloy was supplied by ALTO PARTS. For welding, the aluminium samples were sanded to remove any oxide that might interfere with welding.

3.1 METHODS

3.2.1. SAMPLE PREPARATION

The procedure consisted of oxyacetylene welding PEI glass fibre specimens with AA2024 aluminium alloy standardised for Lap Shear, where the regulated neutral flame is focused directly on the specimen.

Before welding, the area to be welded was sanded to remove the layer of resin or oxide formed from the material as received, and to make it easier to bring the atoms of the structures to be joined closer together.

In order to be able to carry out the welding, two refractory devices were developed with two circular holes so that the flame could be focused on the area of the specimens to be welded, in which the flame comes into direct contact with the aluminium specimen and transfers heat to the PEI specimen. The protection of the sample consisted of adapting this refractory brick, where on both sides an opening was made to fit the aluminium and the sample and a hole was drilled to focus the flame directly on the area where the aluminium was joined, which concentrated the heat so that during welding there was no loss of heat to the environment, improving welding. The refractory device played an essential role in the procedure, because as well as holding the two specimens in the correct position, it protects the part that will not be welded from the flame so as not to degrade the specimen, as well as providing the necessary pressure to allow them to be joined. Figure 6 illustrates the device used.

Figure 6 - Refractory device for soldering.
Source: Author.

After welding, the 8 samples were subjected to the Lap Shear test to obtain stress values (MPa) and then samples were taken from the welded region of the PEI for embedding in resin and microscopic analysis.

3.2.2. WELDING THE SAMPLES

3.2.2.1. conditions and welding procedure

The welding conditions analysed were the height between the torch nozzle and the workpiece, and the time the flame was incident on the aluminium sample. The parameters used were a maximum time of 60 seconds and a minimum of 10 seconds and a maximum distance of 75 mm and a minimum of 30 mm. It was established that within these parameters it was possible to weld and obtain satisfactory results.

Another important factor that changed in the process was the type of flame. Because the carburising flame contains more acetylene than oxygen, it releases a type of soot that forms an oxide layer on the conductive material, which makes it difficult for heat to pass through to the area to be melted. Because of this, the most suitable flame was the neutral flame, which has a higher temperature than the fuel flame, which reduces the welding time and does not release this soot. To adjust the neutral flame, a flow rate of 1 m^3 /h was used for both gases and the neutral plume was adjusted on the torch valves, as shown in Figure 7.

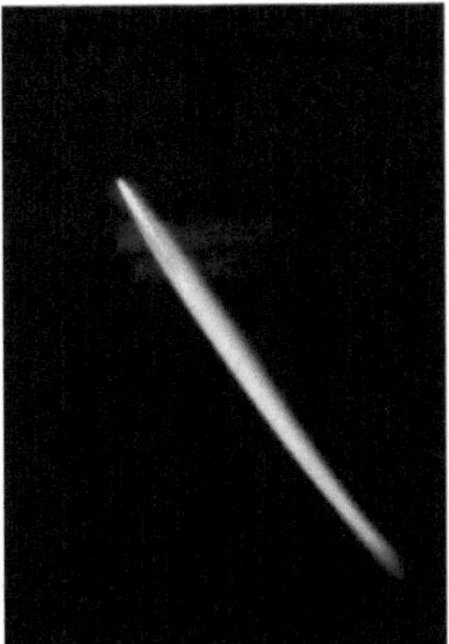

Figure 7-Regulation of the neutral oxyacetylene flame (bright white dart and elongated plume).

Source: Author

Eight specimens were welded to obtain four joint samples. To do this, the torch was fitted to a steel arm that could be moved vertically and horizontally, with the torch nozzle fixed to the screws on the arm itself, as shown in Figure 8.

Figure 8- System used for Oxy-acetylene welding of PEI composite.

Source: Author

3.2.3. LAP SHEAR TEST

The purpose of the Lap Shear test is to quantify the efficiency of welded surfaces of composite materials. It consists of stressing the welded specimen longitudinally until it breaks. The test generally follows the ASTM D 1002 standard, which standardises the dimensions of the specimen at 100 mm long and 25 mm wide and the welding area at 25 mm^2 .

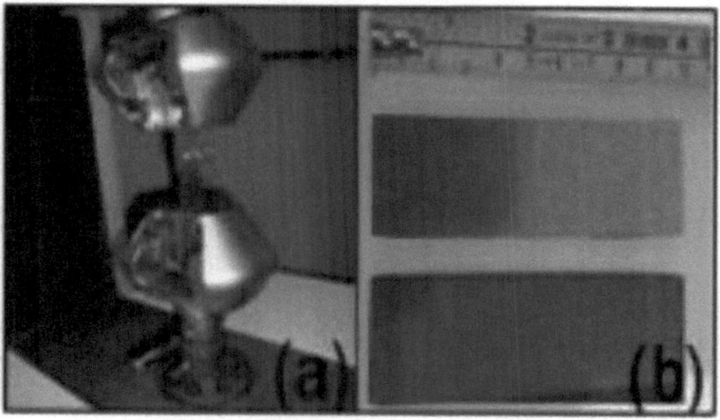

Figure 9 - Representative diagram of the Lap Shear welded specimen.

Source: (ABRAHÃO, 2015).

3.4 OPTICAL MICROSCOPY

To characterise the laminates and assess corrosion, microscopic analysis of the specimens was carried out in the metallography laboratory at FATEC in Pindamonhangaba, using an OLIMPUS optical microscope at 50x magnification, as shown in Figure 10. After preparing the samples, microscopic analyses were carried out to characterise the area where the welding had taken place, using an optical microscope with 50x magnification.

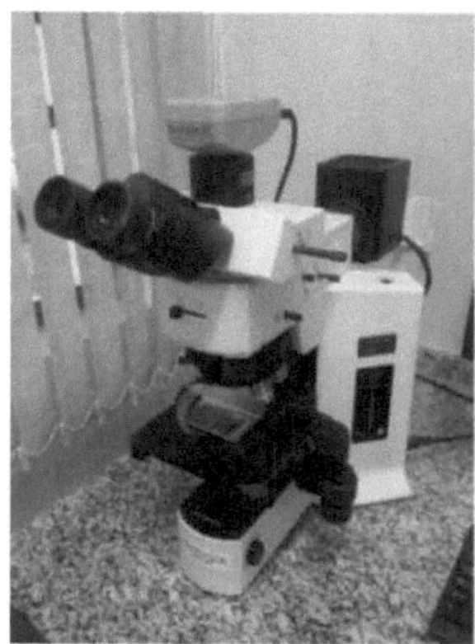
Figura 10 - Optical microscope (Faculdade de Tecnologia de Pindamonhangaba-FATEC)

Source: the Author

10.4.1. Sample preparation

10.4.1.1. PEI fibreglass laminates

To carry out the microscopy, the largest and smallest Lap Shear samples were cut and embedded in RPI-type acrylic resin and left to cure for 24 hours, then sanded with 320, 1200 and 2000 grit wet sandpaper. The samples were then polished with 6 pm diamond paste. Figure 11 shows the samples embedded in acrylic resin.

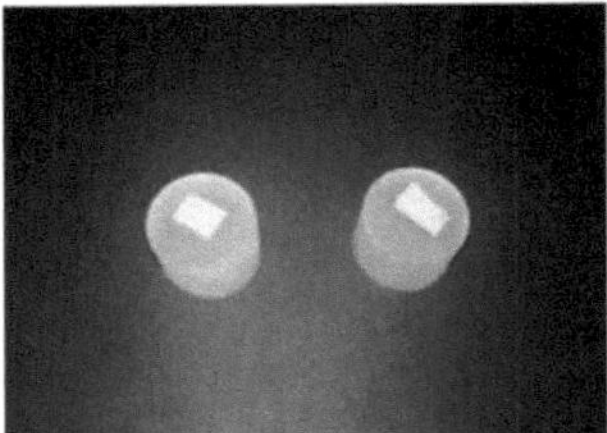

Figura 11 - Samples embedded in acrylic for microscopic analysis.

Source: Author.

3.4.1.2. AA 2024 alloy samples

For the aluminium samples, the conventional procedure of hot-dipping, sanding and polishing was used. A 0.5% hydrofluoric acid solution was used for the chemical attack.

3.5. SEM- Scanning Electron Microscopy

For a more detailed analysis of the fracture in the welded region resulting from the Lap Shear test, images of uncoated samples of the three composites studied were evaluated using the Zeiss Evo LS-15 environmental SEM (Scanning Electron Microscope), using the extended pressure (EP) mode, where the internal pressure of the chamber was adjusted according to each type of sample, between 30 and 100 Pa, in order to compensate for the electrostatic charging effect ("charge-up") by positive nitrogen ions, due to the low conductivity of the samples. The gas used in the chamber was grade 5.0 nitrogen (99.999% purity). Backscattered electron (BSE) signals were sampled using the Zeiss 4Q-BSD detector, a four-quadrant semiconductor, and backscattered electrons using the Zeiss VPSE-G3 detector, specific for pressures up to

400 Pa, which collects photons generated by the breakdown of N2 molecules by secondary electrons that emerge from the sample.

3.6. Energy Dispersive Microanalysis

The energy dispersive microanalysis test was carried out on welded specimens in order to identify the peaks of the elements contained in the composite and on welded samples to analyse how the metal mesh and the welding process affected the polymer matrix. For this purpose, the Oxford Instruments Inca x-act energy dispersive X-ray microanalysis detector mounted on the Zeiss EVO LS-15 scanning electron microscope and controlled by the Inca Energy 250 system, available from the Materials and Technology Department at UNESP-Guaratinguetá, was used.

CHAPTER 4

RESULTS AND DISCUSSIONS

4.1. Morphological characterisation of the materials as received

Therefore, during the development of this work, a careful preliminary microstructural analysis of the composites as received was carried out by optical microscopy using previously polished samples of the PEI/glass fibre laminates and AA Aluminium 2024 alloy before the welding process was carried out. Figures 12 and 13 show the cross-sectional images of the composite laminate samples obtained by optical microscopy, which mainly show the compaction of the fibres with the polymer.

4.1.1. Laminates

The use of the optical microscopy technique to assess the quality of composites has many objectives that can be mentioned, such as: determining the direction and propagation of cracks; assessing the distribution of the homogeneity of fibrous reinforcements in the matrix; checking the degree of impregnation of the fibres by the matrix as well as assessing the fibre/matrix content using programmes dedicated to this purpose (MARINUCCI, 2011).

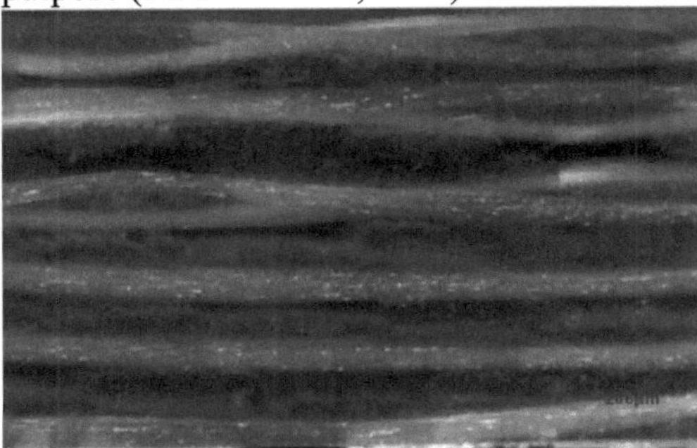

Figure 12-Image of the cross-section of the PEI/glass fibre composite sample at 50 X magnification.

Source: The author.

4.4.2. AA aluminium alloy 2024

The appearance of the aluminium alloy as received is uniform, showing some differences in the phases present due to the chemical composition. Figure 13 shows the image revealed after chemical attack with hydrofluoric acid solution.

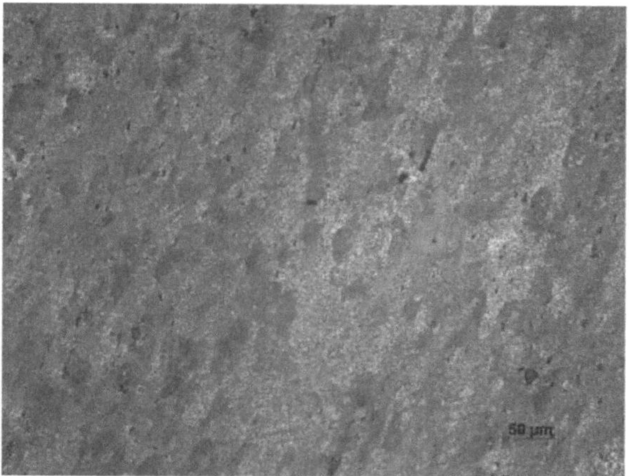

Figure 13. Image of the Aluminium 2024 alloy sample as received, at 50 X magnification.

Source: The author.

4.2. DETERMINING THE INITIAL PARAMETERS FOR WELDING

After developing the welding system, tests were carried out to determine the value ranges and also to study the main variables to be used later. These tests measured the temperature and, above all, whether the material had been welded without degradation, based on the information on the material's glass transition temperature and degradation. With regard to determining the parameters for the materials studied, the first consideration was to visually analyse the welded joints obtained, as shown in Table 1.

Table 1 - Preliminary analysis with the development of welding parameters.

Sample	Time (s)	Distance (mm)	Results
01	60	30	No welding, degradation
02	30	30	Welded out of alignment
03	30	30	Welded without degradation
04	60	30	Welded without degradation
05	60	50	Welded, little degradation
06	60	50	Welded with little degradation
07	60	75	Welded with little degradation
08	10	20	Welded with degradation

Source: Author.

By analysing the results, the welding temperature range in which the composite will be exposed and the minimum and maximum values of the process variables are known, showing that the composite will reach the softening temperature and will be below the degradation temperature. Based on the above results, Lap Shear tests were carried out to check the mechanical strength of the welded joints.

4.3. STUDY OF WELDING CONDITIONS USING THE LAP SHEAR TEST

Subsequently, for a quantitative analysis, the more conventional Lap Shear mechanical test for analysing joints of advanced materials was carried out. The previous tests are now evaluated in terms of mechanical strength, as shown in Table 3 on the next page.

Table 2 - Lap Shear results.

Sample	Tension (MPa}
01	Broke without an MPa value
02	Broke without an MPa value
03	Broke without showing an MPa value
04	141 Mpa
05	Broke without an MPa value
06	6; 23Mpa
07	1.38MPa
06	1.07 MPa

Source: Author.

As a way of analysing the results, the stress values obtained in the Lap Shear mechanical test were the main considerations, where sample 06, although it suffered a little degradation, was still the best result, and the worst result (lowest stress value obtained) was from sample 08, which, in addition to suffering degradation due to the short distance (20 mm), the welding was also fragile due to the short flame incidence time (10 seconds). Figures 6 and 7 show the difference between the best and worst results.

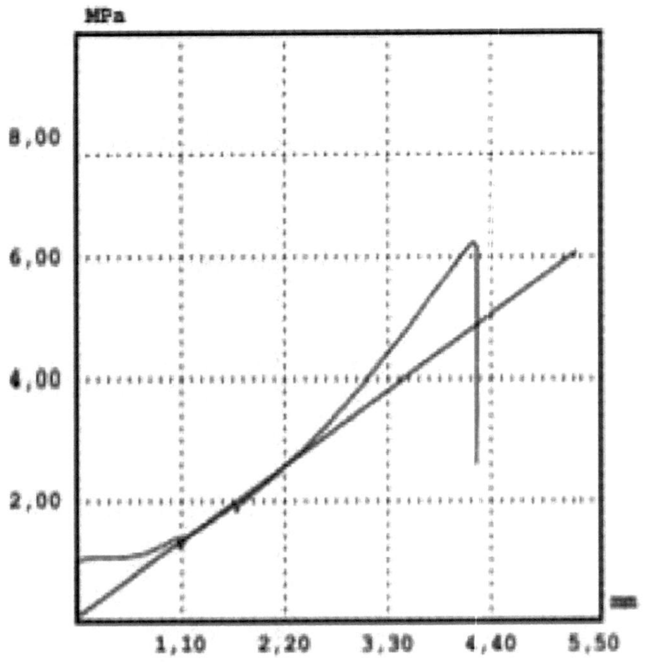

FORÇA MÁXIMA N	LIMITE DE RESISTÊNCIA MPa	LIMITE DE ESCOAMENTO MPa	RELAÇÃO LR/LE	ALONGAMENTO %	ALONGAMENTO LINEAR mm	ALONGAMENTO (%) CARGA MÁXIMA %	ALONG. LINEAR CARGA MÁXIMA mm	LARGURA mm	ESPESSURA mm
3893,30	6,23	0,00	0,00	8,52	4,260	8,48	4,240	25,000	25,0000

Figure 14 - Result of the highest stress value obtained by the *Lap Shear* test (Sample 06).
Source: Author

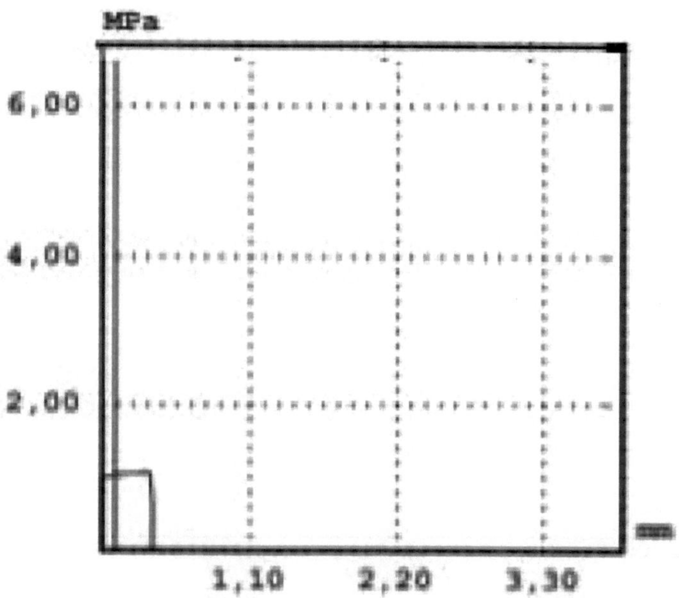

Amostra No.	FORÇA MÁXIMA N	LIMITE DE RESISTÊNCIA MPa	LIMITE DE ESCOAMENTO MPa	RELAÇÃO LR/LE	ALONGAMENTO %	ALONGAMENTO LINEAR mm	ALONGAMENTO (%) CARGA MÁXIMA %	ALONG. LINEAR CARGA MÁXIMA mm	LARGURA mm	ESPESSURA mm
1	666,86	1,07	0,00	0,00	0,76	0,380	0,72	0,360	25,000	25,0000

Figure 15 - Result of the lowest stress value obtained by the *Lap Shear* test (Sample 08).
Source: Author

The graphs generated after the Lap Shear test show the difference in the results obtained with oxyacetylene welding. The first graph (Figure 14) shows a well-defined curve, with elongation of the material before failure, which obtained a mechanical strength value of 6.23 MPa. On the other hand, the graph of the worst Lap Shear sample (Figure 15) makes it clear that the welded joint was very weakened due to the short time the flame was present and due to the very close flame that degraded part of the PEI-Fibreglass, obtaining a value of 1.07 Mpa.

4.3. ANALYSIS OF THE JOINT OBTAINED BY OPTICAL MICROSCOPY

For a better visual analysis, metallographic analyses of the region where the welding took place were carried out on both the PEI-Fibreglass samples and the AA2024 aluminium alloy samples. According to the metallography on the PEI-Fibreglass, the presence of voids in the structure of the PEI can be seen, which are due to the breaking of the weld by means of the Lap Shear test. These voids indicate the places where the fibreglass adhered most to the AA2024 alloy. In addition, the residual AA2024 alloy is noticeable in the vicinity of the voids, as we can see spots that are slightly lighter than the PEI structure. In addition, there are parts of the PEI that have not adhered and have remained intact, with no voids or alterations. This can be seen in Figure 16 a and b, where the weld was analysed.

Figure 16. Welded joint obtained before the LAP SHEAR test.

Figure 9 illustrates the part of the composite after breaking by the mechanical test. As can be seen, no pores were formed as a result of adhesion with the AA2024, leaving a dark colour where the PEI polymer matrix detached.

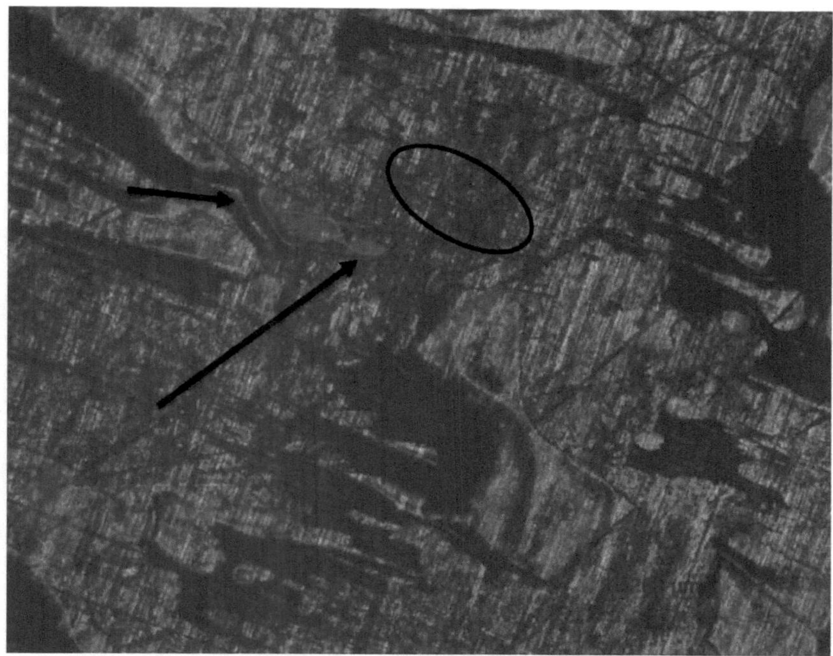

Figure 17 - Image of the PEI-Fibreglass welded area after the *Lap Shear* test at 50 X magnification
Source: Author.

When analysing the microscopic image of the AA2024 specimen that showed the best results, we can see the presence of voids resulting from parts of the alloy that have come off due to the Lap Shear mechanical test; these regions are the ones that adhered the most during the welding process. We can also see parts of lighter spots that are residual PEI still present in the 2024 alloy. In addition to the parts of the aluminium alloy that remain intact. These regions are exemplified by Figure 19 on the next page.

Figure 18 - Microscopy of the welded area of the Al AA2024 alloy after the Lap Shear test .

Source: Author.

Figure 19 below shows the metallography of the sample with the lowest tensile strength. The action of degradation is visible, which has left the image very dark, and there are no voids present as a result of good adhesion. We can only see a large linear spot where there was low-intensity adhesion, but which was enough to hold the weld, although it showed a low Lap Shear result.

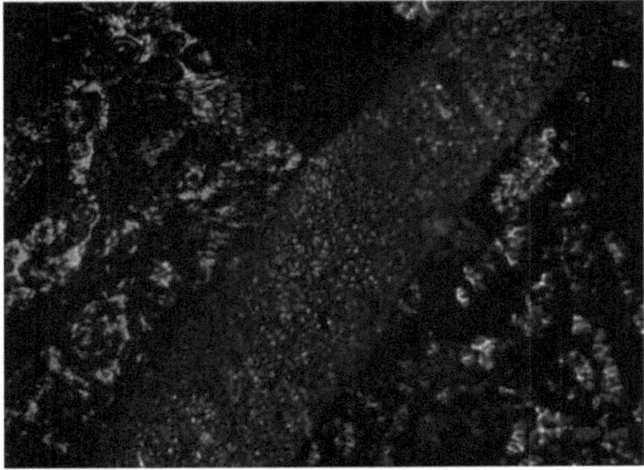

Figure 19 - Microscopy of the welded area of the Al AA2024 alloy after the Lap Shear test (Lowest value)

Source: Author.

4.4. ELECTRONIC MICROSCOPY and EDS TESTING

Figure 20 shows the diagram obtained from the EDS analysis, revealing the chemical components of the aluminium sample in the weld region. This evaluation was carried out on the sample after it had been subjected to the Lap Shear test, showing the contact between the aluminium and the polymer (matrix) and the fibres. The region analysed is contained in the rectangle in Figure 20.

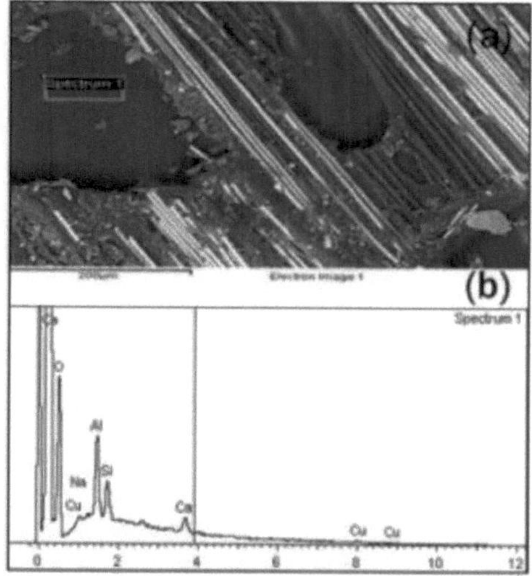

Figure 20- (a) EDS analysis of the interaction region between the aluminium and the laminate of the welded aluminium sample; (b) elemental spectrum of the analysed region.

Table 3. Chemical composition of the broken part of the sample.

Element	% mass	% atomic mass
O	72,47	82,91
In	1,84	1,47
Al	13,14	8,92
Yes	6,66	4,34
Ca	4,01	1,83
Ass	1,88	0,54

The results in Table 3 show the same elements found in the interaction between aluminium 2024 and the PEI/glass fibre laminate. In addition, the mass percentages of oxygen, aluminium and silicon are the highest in the selected aluminium region. The oxygen comes from the polymer and the welding environment, the aluminium comes from the sample and the silicon comes from the glass fibres. All the chemical elements identified in this study are constituents of the welded joint and the materials of origin, and there is no formation of a new constituent that could interfere with the quality of the weld.

CHAPTER 5

CONCLUSION

According to the welding tests carried out, the definition of a neutral flame, a refractory device for welding, variations in flame distance and heat incidence time. Another important factor for welding was sanding the area of the AA2024 that was to be welded, as this removes a layer of surface resin present on the alloy.

Subsequently, the Lap Shear mechanical test values obtained made it clear that it is possible to weld these dissimilar materials with satisfactory mechanical strength results, although this is still an experimental process. This makes it possible to use PEI-glass fibre with AA2024 aluminium alloy in aeronautics, among other areas of application.

REFERENCES

ABRAHÃO, A. B. R. M. Optimisation of the electrical resistance welding process in continuous pei/fibre composites for aeronautical applications. Doctoral thesis. UNESP. Guaratinguetá. 2015.

ABREU, C. P. Characterisation of the reactivity of friction stir welded (FSW) AA2024-T3 and AA7475-T651 aluminium alloys. Doctoral thesis. UNESP. São Paulo. 2016.

ADHIKARI, B; MAJUMDAR, S. Polymer in Sensor Applications. Prog.Polym.Sci, 2004, p. 699 - 766.

AGEORGES, C.; YE, L. Resistance Welding of Metal/Thermoplastic Composite Joints. Journal of Thermoplastic Composite Materials, v. 14, n. 6, p. 449-475, 2001.

AGEORGES, C.; YE, L.; HOU, M. Experimental investigation of the resistance welding for thermoplastic-matrix composites. Part I: heating element and heat transfer. Composites Science and Technology, v. 60, n. 7, p. 1027-1039, 2000.

BRAZILIAN ALUMINIUM ASSOCIATION - ABAL. Fundamentals and Applications of Aluminium. São Paulo: ABAL Publishing House, 2007.

BATISTA, N.L; BOTELHO, E. C. Influence of weathering on the viscoelastic performance of pei/carbon fibre laminates for aerospace applications. CBPol, 10th. UNESP. Guaratinguetá. 2009.

BATES, P.J. Resistance welding of thermoplastic composites-an overview. Composites: Part A, 36, 2005, 39-54.

BARROS, F.B. Preparation, physicochemical characterisation and evaluation of the thermal and mechanical behaviour of poly PET/PEI blends. PhD thesis in physical chemistry, UFSCAR, São Paulo 2007, 216p.

BRACARENSE, A.Q. Oxy-Gas Flame Welding Process - OFW. Belo Horizonte: Editora UFMG, 2000.

CATSMAN, P. Polyetherimide (PEI): Substituting Metal. Germany: Kunststoffe, 2005, p.143-147.

COUTO, A. A; REIS, D. A; JUNIOR N.I. et al. Study of the fatigue mechanical properties of artificially aged AA2024 aluminium alloy for aeronautical applications. 5th National Aluminium Congress. São Paulo. 2012.

DA COSTA, A.P. **Effect of environmental conditioning on welded PPS/Continuous Fibre composites.** Guaratinguetá: Universidade Estadual Paulista, Faculdade de Engenharia de Guaratinguetá, 2011.

DA SILVA, I. L. A. Properties and structure of polymer composites reinforced with continuous jute fibres. Doctoral thesis. UENF. Rio de Janeiro. 2014.

DUBÉ, M. et al. Current leakage prevention in resistance welding of carbon fibre reinforced thermoplastics. **Composites Science and Technology**, v. 68, n. 6, p. 1579-1587, 2008.

GOMES, M. A. Mechanical properties of polymer composites reinforced with pineapple leaf fibres (palf). PhD thesis. UENF. Rio de Janeiro. 2015.

KHAN, S. S. Low cycle lifetime assessment of Al2024 alloys. Doctoral thesis. Technische Universit at Dortmund. Germany. 2011.

MARQUES, P.V.; MODENESI,P.J.; BRACARENSE, A.Q. **Welding Fundamentals and Technology.** Editora UFMG, Belo Horizonte 2009.

MARTINS, J. V. Mathematical modelling of mechanical properties of friction composite materials. Undergraduate Thesis. UFSC. Florianópolis. 2007.

MODENESI, P. J. Soldagem I Introduction to Welding Processes. UFMG. Minas Gerais. 2000.

NETO, L.P.O. The application of polyamide 6.6 in high performance parts for the furniture market. São Paulo: Faculdade de Tecnologia da Zona Leste, 2009.

OLIVEIRA, G. H; BOTELHO, E. C. Evaluation of the Fatigue Resistance of Carbon Fibre/PEI Composites with Applications in the Aerospace Industry. CECEMM, 9TH. UFSC. Florianópolis. 2007.

OLIVEIRA, G. H.; GUIMARÃES, V. A.; BOTELHO, E. C. Influence of temperature on the mechanical performance of PEI/glass fibre composites. Polímeros, v. 19, n. 4, p. 305-312, 2009.

PANNEERSELVAM, K.; ARAVINDAN, S.; NOORUL HAQ, **A. Study on resistance welding of glass fibre reinforced thermoplastic composites.** Materials & Design, v. 41, p. 453^59, 2012.

SILVA, H. S. P. Development of polymer composites with curauá fibres and glass fibre hybrids. Master's Thesis. UFRGS. Porto Alegre. 2010.

SOUZA, E.W. MARINUCCI, G. **Study for the manufacture of automotive reflectors using a thermosetting composite material and a thermoplastic material.** São Paulo: Institute for Energy and Nuclear Research, University of São

Paulo, 2010.

STAVROV, D.; BERSEE, H. E. N. **Resistance welding of thermoplastic composites-an overview.** Composites Part A: Applied Science and Manufacturing, v. 36, n. 1, p. 39-54, 2005.

YOUSEFPOUR, A.; HOJJATI, M.; IMMARIGEON, J.-P. **Fusion Bonding/Welding of Thermoplastic Composites.** Journal of Thermoplastic Composite Materials, v. 17, n. 4, p. 303-341, 2004.

VILLANI, P.M.; MODENESI, P.J.; BRACARENSE, A.Q. Welding Fundamentals and Technology. Belo Horizonte: Editora UFMG, 2009.

WAINER, E.; BRANDI, S.D.; MELLO, F.D.H. Soldagem: processo e metalurgia. São Paulo: Editora Edgard Bluncher LTDA, 1992.

WISE, R.J. **Thermal Welding of polymers.** Cambridge, Woodhead Publishing Limited, 1999.

ZANATTA, R. Composite materials in aviation. Article. Aviation.org. 2012.

Printed by Books on Demand GmbH, Norderstedt / Germany